ICE STORMS:
CAUSES AND EFFECTS

by Anne Winch

12
STORY LIBRARY
MORE TO EXPLORE

www.12StoryLibrary.com

12-Story Library is an imprint of Bookstaves.

Developed and produced for 12-Story Library by Focus Strategic Communications Inc.

Library of Congress Cataloging-in-Publication Data
Names: Winch, Anne, author.
Title: Ice storms : causes and effects / by Anne Winch.
Description: Mankato, Minnesota : 12-Story Library, [2022] | Series: Wild weather | Includes bibliographical references and index. | Audience: Ages 10–13 | Audience: Grades 4–6
Identifiers: LCCN 2020018036 (print) | LCCN 2020018037 (ebook) | ISBN 9781645821519 (library binding) | ISBN 9781645821892 (paperback) | ISBN 9781645822240 (pdf)
Subjects: LCSH: Ice storms—Juvenile literature.
Classification: LCC QC926.37 .W56 2022 (print) | LCC QC926.37 (ebook) | DDC 551.55/6—dc23
LC record available at https://lccn.loc.gov/2020018036
LC ebook record available at https://lccn.loc.gov/2020018037

Photographs ©: Bob Hilscher/Shutterstock.com, cover, 1; The Old Major/Shutterstock.com, 4; Douglas Knisely/CC2.0, 5; Jake N./CC2.5, 5; Irina Mos/Shutterstock.com, 6; White_Fox/Shutterstock.com, 7; R. Rose/Shutterstock.com, 7; Dmitri Ma/Shutterstock.com, 8; Sundry Photography/Shutterstock.com, 9; Joey Thebeau/Shutterstock.com, 9; NOAA/Dan Borsum, 10; Kirk Hewlett/Shutterstock.com, 11; NataBystrova/Shutterstock.com, 11; Jim Lane/Alamy, 12; rthoma/Shutterstock.com, 13; Lindsay Rustad/U.S. Forest Service Northern Research Station, 13; Muskoka Stock Photos/Shutterstock.com, 14; NOAA Weather in Focus Photo Contest 2015/Mike Shelby, 14; ValeStock/Shutterstock.com, 15; ANGHI/Shutterstock.com, 15; Uti Palou/Shutterstock.com, 16; Tony Campbell/Shutterstock.com, 17; Zuzule/Shutterstock.com, 17; Lodo27/CC3.0. 18; Design Pics Inc./Alamy, 18; Megapress/Alamy, 19; Shanshan0312/Shutterstock.com, 19; MAHATHIR MOHD YASIN/Shutterstock.com, 20; John Penezic/Shutterstock.com, 21; FZappa20/YouTube.com, 21; Wanda Olsen/Shutterstock.com, 22; Joy Brown/Shutterstock.com, 23; EvGavrilov/Shutterstock.com, 23; Video131/Shutterstock.com, 24; Deb/CC2.0, 25; Bert Folsom/Alamy, 25; NASA/Mike Studinger, 26; Bernhard Staehli/Shutterstock.com, 27; Andreas Weith/CC4.0, 27; Africa Studio/Shutterstock.com, 28; Haoka/Shutterstock.com, 29; Radovan1/Shutterstock.com, 29

About the Cover

Trees damaged from the effects of an ice storm.

Access free, up-to-date content on this topic plus a full digital version of this book. Scan the QR code on page 31 or use your school's login at 12StoryLibrary.com.

Table of Contents

1

Dangerous Ice Storms

Ice storms are caused by freezing rain. Raindrops land on a surface that is below freezing temperature. The raindrops freeze into a thin sheet of ice. Sometimes this smooth sheet of ice is called glaze ice. The ice forms on trees, cars, roads, and sidewalks. It is very dangerous for walking or driving.

Imagine candy coated in a hard, crunchy sugar shell. Places struck by ice storms look like candy lands. Ice glazes everything. Roads. Trees.

Cars. Buildings. Sidewalks. This hard, slick shell makes walking or driving dangerous.

Ice storms are gentle storms. Freezing rain drips or tip-taps. Danger comes after the storm. Ice cannot be plowed away like snow. Icicles fall like daggers.

THINK ABOUT IT

How would you describe an ice storm?

90

Top speed in miles per hour (145 km/h) of large falling icicles

- Large icicles can smash the ground with 1,000 pounds (454 kg) of force.
- A quarter inch (6 mm) ice layer is all it takes to have an official ice storm.
- In some parts of the US, an ice storm is called a silver thaw. It looks like a silver coating of ice on the surface of things.

2

How Ice Storms Form

Fly 6,500 feet (1,980 m) into Earth's atmosphere. There, you'll see where ice storms begin. Snow falls from this height. If it passes through a warm air current, it melts into rain. If the temperatures near Earth's surface are below freezing, the rain supercools. That is, it hardens into ice.

500

Weight in pounds (225 kg) that an extra half inch (10 mm) of ice adds to power lines

- The excess weight stresses and breaks power lines.
- Ice can increase the weight of a tree branch by 30 times.
- Freezing rain is the most dangerous form of precipitation.

Freezing rain forms a hard, icy layer on surfaces. It can also dangle as icicles. Ice is dangerous because it is slippery and heavy.

BLACK ICE

Freezing rain causes more road accidents than other types of weather. Black ice is a thin layer of clear ice on paved roads. The ice is so thin, it blends in with the black pavement. Icy roads cause over 150,000 accidents each year.

3

Predicting an Ice Storm is Tricky

Winter brings many storms. Forecasting, or predicting, these storms saves lives. Predicting an ice storm is tricky.

Meteorologists use computer models to predict the weather. A model uses math to compare past weather trends. The trends can predict the future. Models rely on data collected by weather tools. The information is entered into a supercomputer. The computer fits the data into different weather models. A model may show that an ice storm is possible.

NEXRAD radar station in northern California.

In the US, the National Weather Service (NWS) collects the data. The data come from many sources. Radar tracks precipitation, or rainfall. Weather balloons collect data high in the atmosphere. Satellites monitor weather patterns.

80,000 to 120,000

Maximum height in feet (24,400 to 36,600 m) that weather balloons reach

- Ninety two weather stations in the US release weather balloons.
- The American weather radar system is called NEXRAD. NEXRAD stands for Next Generation Radar.
- The Department of Defense operates some of the NEXRAD systems.

I SPY ICE SPY

Ice storms happen when the raindrops move through air layers at different temperatures. Small changes in temperature can change rain to freezing rain, sleet, or snow. These changes make ice storms very hard to predict.

4 Alerts, Warnings, Watches, Oh My!

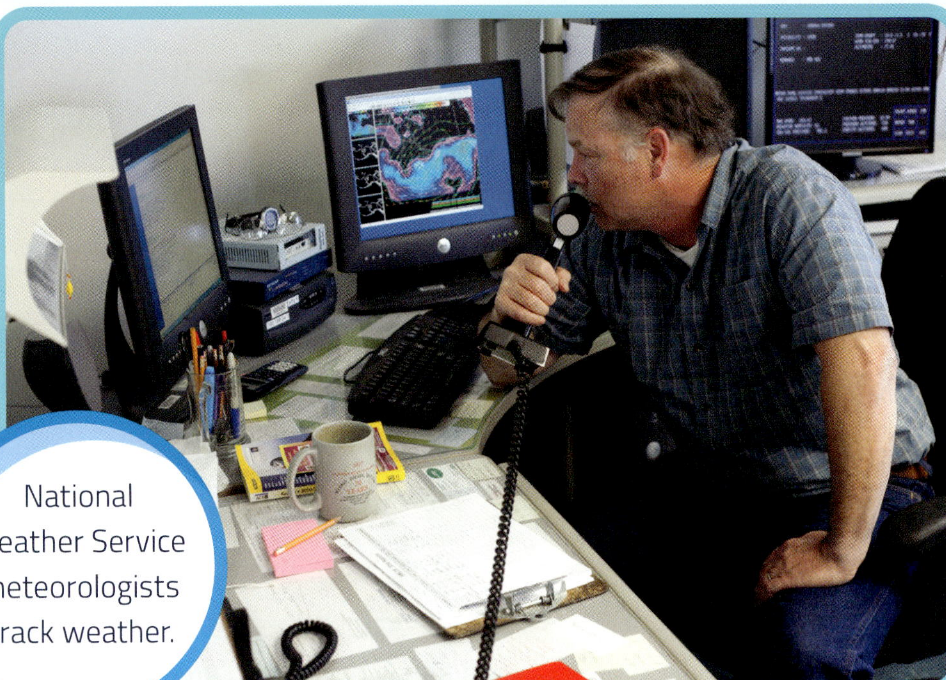

National Weather Service meteorologists track weather.

Is an ice storm coming? The National Weather Service (NWS) provides weather alerts. The NWS sends alerts to organizations like radio stations. They share the alerts with the public.

The NWS issues a *winter weather watch* 24 to 72 hours ahead of any winter storm. An *ice storm warning* is issued when an ice storm is very likely to occur in the next 12 to 36 hours. Sometimes the storm is already in progress when the warning comes.

A *winter weather advisory* means stormy weather is on the way or has arrived. It is

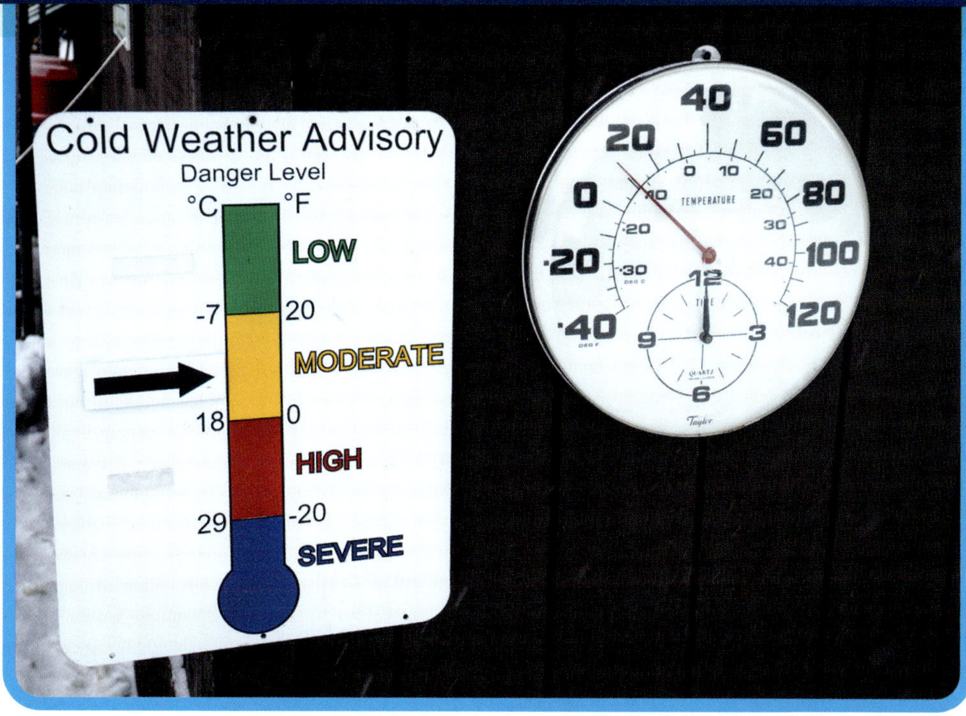

less serious than a warning. Advisories apply to snow, freezing rain, or sleet.

Alerts help cities and states prepare for ice storms. Officials ask people to stay home and off the roads. Electricity companies prepare for fallen wires and power outages.

0.25 to 0.50

Amount of ice build-up in inches (0.64 to 1.27 cm) predicted before an ice storm warning is issued

- The amount of ice needed for a warning depends on where the storm is happening.
- The NWS releases *heavy freezing spray* warnings when freezing spray causes a lot of ice on a boat.
- Schools may be closed if an ice storm is coming.

5

Measuring an Ice Storm Isn't Easy

Scientists use special measurements to gage the size of a storm. Ice storms are measured by the Sperry-Piltz Ice Accumulation Index, or SPIA. It is named after the two men who invented it, Sidney Sperry and Steven Piltz.

The index measures three things. First, the amount of ice that freezes on surfaces. The ice indicates how much rain has fallen. For instance, 10,600 gallons (40,125 l) of water covers an ice rink with 1 inch (2.5 cm) of ice.

80

Number of hours freezing rain fell in the biggest ice storm in Maine's history

- An ice storm warning is issued when large amounts of ice might stick to surfaces.
- The cold air under a bridge causes its surface to freeze faster than roadways.
- The slickest ice forms when temperatures reach 19° Fahrenheit (–7° C).

The SPIA also measures the wind speeds during the storm. Lastly, it tracks the amount of damage caused. Damages are rated on a scale from zero to five. Zero is the least amount of damage, and five is the most damage. This scale warns people about ice storms. It helps people prepare for the storm. When it comes to saving lives, a little warning goes a long way.

THINK ABOUT IT

Scientists create their own ice storms. They pump water from streams into freezing air. What might they learn from this process?

6

Ice Storms Are a Terrible Beauty

After an ice storm, the world resembles the Ice Queen's palace. The tree branches sparkle and shine with their new coating of ice. Icicles hang like white ribbons. However, all that ice can cause a lot of damage.

Ice is heavy. The extra weight snaps off tree branches. On the ground, those branches can block roads, crush vehicles, and injure people. Some trees split down

Ice storms damage many trees each year.

the trunk. Cleaning up trees after an ice storm is a big job.

In an ice storm, trees lose many branches all at once. Without branches, trees struggle to survive. They may not have enough leaves to get energy from the sun. Harmful insects burrow inside and kill the trees.

90 to 95

Percent of ice-damaged forest trees that were still alive five years after a major ice storm in New England in 1998

- Trees regrew their crowns (the top part) in three to five years.
- Evergreens, like spruce, are more resistant to ice breakage than deciduous trees (trees with leaves).
- Homeowners and city governments can plant and care for ice-resistant trees.

7

Ice Storms Harm Animals

Wet or icy fur can make it hard for animals to stay warm.

Freezing rain affects animals in many ways. First, it can cover their bodies in ice. Birds cannot fly with heavy ice on their wings and tails. Ice on animal fur makes it harder to keep warm.

Ice forms a hard cap on top of snow. This cap traps small mammals, such as mice.

Minus 40°

Temperature in degrees Fahrenheit (– 40°C) at which snowy owls can still easily survive

- Branches bent under heavy ice provide shelter for some animals.
- Animals cannot hunt for food during an ice storm. They live off their body's stored fat.
- The feet of small birds sometimes freeze to branches.

They cannot get fresh air. Predators cannot dig under the snow to hunt for prey. Deer cannot eat grass locked under the icy layer. Many animals go hungry.

KEEP YOUR DOG AT HOME

Ice storms can be very dangerous for pets as well as wildlife. Dogs should be leashed. The scents they use to navigate disappear on ice. Ice-melting salts and the antifreeze fluids in cars can poison dogs.

It can be dangerous for pets to try and walk on ice.

8

Ice Storms Happen Around the World

In 2010, many parts of Moscow lost power after a huge ice storm.

A lot of ice storms occur in North America. However, ice storms strike other places around the world, too. In 2010, a serious ice storm cut off power to hundreds of towns around Russia's capital, Moscow. Four inches (10 cm) of ice crusted over Slovenia in 2014. Denmark woke up to freezing rain in February 2017.

Climate change may increase ice storms. Scientists believe that as the Earth gets warmer, ice storms will become more

Huge power towers toppled under the weight of ice during the 1998 storm.

17 million

Number of acres (6,879,656 hectares) of forest damaged across New England during the ice storm of 1998

- In 1961, an ice storm in Idaho produced 8 inches (20 cm) of ice.
- Two ice storms hit Atlanta, Georgia, just before it hosted the Super Bowl in 2000.
- Even sunny Northern Florida sometimes gets freezing rain.

serious. Melting polar ice causes more violent rain and windstorms. Adding more rain or wind to an ice storm only makes it more dangerous.

PANCAKE PANIC

Ice storms can threaten your favorite breakfast foods. Parts of North America are famous for making maple syrup. One source estimated that 65 percent of sugar maple trees were damaged in a 1998 ice storm. Pancakes went without syrup while the maple syrup industry recovered.

9

Cities Fall Victim to Ice

This car became encased in a thick layer of ice.

Ice storms can devastate a city or town. Large populations are exposed to the ice and its many dangers. A city also has many vehicles, buildings, and businesses. Airports, trains, and shipping can be interrupted.

Trees and utility poles topple. They can crash into buildings or block streets. Downed wires

mean no heat and electricity. Sometimes people use dangerous heat sources to keep warm. As a result, they may suffer from gas poisoning. Before paved roads, ice storms blocked travelers less often. Ice cannot form a solid layer on dirt roads. However, ice is a hazard on smooth, paved roads. Many vehicles crash or are abandoned because they cannot drive on the slick surface.

4.3 million

people in Texas who lost power in a February 2021 ice storm

- The 2010 Moscow ice storm stranded tens of thousands of people in airports and on trains.
- Over 2,000 vehicles were abandoned on a highway after a 2014 ice storm in Atlanta.
- The February 2021 storm caused at least $195 million in damages.

Cities like Atlanta are not used to ice storms.

10

Ice Plays Havoc on Farms

Rural ice storms can be dangerous. Rural areas are very remote and isolated. A person in trouble may be far from help.

Power outages cause many problems on a farm. Feeding or milking machines won't work. Buildings have no lights. Electric water pumps stop, and

It is difficult for livestock to move in icy fields after an ice storm.

water may be scarce. Repairs are delayed because repairmen must travel long distances.

Despite the ice, livestock need to be fed and watered. Farmers risk their own lives. They trek across the icy ground to feed the animals. They cannot stay indoors and wait for the ice to melt.

Barn roofs can collapse under the weight of the ice. As a result, animals lose their shelter. Worse, they can be injured or killed. Fields covered with ice send animals slipping and falling.

2

The number of times most dairy cows are milked each day

- If cows are not milked during an ice storm, they may get sick.
- Hypothermia (very low body temperature) is a deadly problem during an ice storm.
- Icicles and ice blobs on power lines jiggle and "gallop" in the wind. This stress breaks the lines.

11 The Incredible Ice Storm of 1998

In January 1998, a series of ice storms slammed parts of upstate New York, Vermont, New Hampshire, Maine, and Canada. The storm struck a narrow band of land. Even so, the impact was huge.

The storm began when warm air from the Gulf of Mexico hit cold air from the Arctic. The warm air released rain that froze when it hit the ground.

Huge steel towers crumpled. Power lines and utility

Over $4 billion

The cost of the great ice storm of 1998

- Dozens of people died in the US and Canada.
- People died from cold, fires, carbon monoxide (gas) poisoning, and injuries.
- In Canada, over 15,000 members of the Armed Forces were deployed to help.

poles collapsed. Millions of people lost power. Some were without power for weeks. Roads closed. Snow, ice, stranded vehicles, and fallen trees blocked routes.

Animals died on farms and in the wild. Millions of trees in forests and orchards were damaged.

MAINE WAS LIKE A GIANT ICE SCULPTURE

In Maine, the icy rain fell for about 50 hours. More than half the residents lost power. Schools were canceled for two weeks. People helped their neighbors. They made sure everyone was safe.

12

Ice Storms and Climate Change

Climate is a weather pattern across a long period of time. Earth's atmosphere impacts the climate. Currently, the atmosphere is warming. The extra heat changes weather patterns. Some bodies of water no longer freeze completely. Glaciers, Arctic ice, and ice caps are melting. More moisture evaporates into the air. This can create more freezing rain. Ice storms may occur more often. They may last longer. The ice layers will be thicker.

When large chunks of ice break off the glacier, it is known as ice calving.

Warmer air also alters the polar jet stream. The jet stream is a blast of cold Arctic air. It mixes with warm air near the equator. The frigid air travels further south. As a result, more extreme weather arises. Ice storms are just one example of extreme weather.

80

Percent of sunlight that sea ice reflects back to space, helping keep Arctic temperatures cool

- When Arctic ice melts, the ocean warms. Dark water absorbs more heat energy.
- Climate change impacts the Rocky Mountains. Millions of trees have died because of wildfires, heat, drought, and insects.
- Winter sports may be lost due to shorter winters and extreme storms.

THINK ABOUT IT

How could extreme storms impact where you live?

Animals can become trapped away from their hunting grounds.

Staying Safe in an Ice Storm

How do you weather an ice storm? Here are some ways to stay warm and safe.

- Stay inside and dress warmly.
- If the power goes off, stay in one room. Close off all the other rooms to save heat.

- Put towels or blankets under doors to keep the heat in.
- Cover the windows. This helps prevent heat loss through the glass.
- Make sure there are working flashlights and extra batteries.
- Keep the fridge closed as much as possible to keep food cold.
- Keep some canned or boxed food on hand.
- Have bottled water available.
- Have a crowbar or ax in case you must chop through ice to exit the house.
- Use a portable generator to keep power in parts of the house.

Glossary

alert
A notification of a possible danger or problem.

atmosphere
The layer or layers of gases surrounding a planet

carbon monoxide
A colorless and odorless gas that is poisonous.

glaze ice
A smooth, transparent coating of ice caused by freezing rain.

meteorologist
A scientist who studies and predicts the weather.

sleet
A mixture of rain and snow frozen into ice.

supercool
Cooling a liquid or gas below its freezing point without freezing it.

translucent
Letting some light through but not completely clear.

utility
A public service like electricity, gas, or water.

weather balloon
A balloon equipped with tools that is sent into the atmosphere to gather information about the weather.

Read More

Black, Vanessa. *Disaster Zone: Ice Storms*. Minneapolis, MN: Jump! Pogo Publishing, 2016.

Cosgrove, Brian. *Weather: Discover the World's Weather from Heat Waves and Droughts to Blizzards and Flood*. New York, NY: DK Children, 2016.

Fleisher, Paul. *Lightning, Hurricanes, and Blizzards: The Science of Storms*. Minneapolis, MN: Lerner Publications, 2010.

Furgang, Kathy. *National Geographic Kids Everything Weather: Facts, Photos, and Fun that Will Blow You Away*. Washington, DC: National Geographic Kids, 2012.

Visit 12StoryLibrary.com

Scan the code or use your school's login at **12StoryLibrary.com** for recent updates about this topic and a full digital version of this book. Enjoy free access to:

- **Digital ebook**
- **Breaking news updates**
- **Live content feeds**
- **Videos, interactive maps, and graphics**
- **Additional web resources**

Note to educators: Visit 12StoryLibrary.com/register to sign up for free premium website access. Enjoy live content plus a full digital version of every 12-Story Library book you own for every student at your school.

Index

About the Author

Anne Winch is an elementary school teacher in Ontario, Canada. She lives with her husband and two Doodle puppies. She is an avid reader, writer, and lifelong learner. She has finally found a way to incorporate all of these passions.